Railways in Profile S…

BRITISH RAI…
ENGINEER'S STOCK
Volume Two

Compiled by G. Gamble

CHEONA PUBLICATIONS

ISBN 1 900298 08 2

Production, design and setting by:
Print Rite, The Willows, School Lane,
Stadhampton, Oxford OX44 7TR

Published by:
Cheona Publications
39 The Avenue, Chinnor,
Oxfordshire OX9 4PD

Printed by Alpha Print, Crawley, Oxon.

The Railways in Profile Series.

No. 1 British Railway Wagons. Opens and Hoppers.
No. 2 Great Western Stations.
No. 3 British Railway Vans.
No. 4 British Railway Wagons. Railtanks.
No. 5 British Railway Brake Vans and Cattle Wagons.
No. 6 British Railway Non-Passenger Rolling Stock.
No. 7 British Railway Wagons. Engineer's Stock - *Volume One*
No. 8 British Railway Engineer's Stock - *Volume Two*

Acknowledgements

Such a work as this relies heavily on the availability of photographs and grateful thanks are extended to all who have helped in this matter. They have been indicated by initials to each plate and are: R.C. - Roger Carpenter, P.C. - Phil Coutanche, B.D. - Brian Daniels, G.G. - Geoff Gamble, S.J. - Steve Jordan, D.L. - David Larkin, J.M./R.C. - Joe Moss (courtesy Roger Carpenter), N. - Nelson collection (courtesy of Colin Judge), F.W.S - 'Tim' Shuttleworth, R.SL. - Roger Silsbury, R.S. - Roger Speller, A.T. - Alan Turner, and H.F.W./R.C. - Hubert Wheeler (courtesy of Roger Carpenter).
Many of the above and other friends have helped by answering questions and supplying further information for which I am most grateful. Too many books have been consulted to give a bibliography here but the various editions of 'Departmental Coaching Stock' published by Platform 5 Publishing Ltd. have frequently been referred to.
Many thanks go to Roger Carpenter, Roger Silsbury, 'Tim' Shuttleworth and Neal Suckling who have produced excellent monochrome prints and also to Colin Judge for his help in the design and production of this book.

Preface

Like Profile No. 7 in the 'Railways in Profile' series, this volume covers more vehicles serving in the Engineering Department. Here the subjects are items of coaching stock along with some vehicles which have been converted to work with them all of which were originally in revenue earning service. Many of these vehicles will make interesting challenges to modellers using proprietary models or kits as starting points for the conversions. Sadly it has not been possible to obtain views of both sides of most of the vehicles. The very specialised items such as de-icers, sandite units, laboratory coaches, track-recording coaches etc. will have to be covered in future volumes.

Introduction

In early BR days, long term railway development and maintenance projects necessitated the staff concerned living close to the sites which were often some distance from their homes. For this purpose staff accommodation vehicles were needed with mess and dormitory facilities, mobile workshops and associated tool and stores vans. With the onset of electrification special overhead line construction and maintenance coaches were also needed. The development of motorways, upgrading of minor roads and the rationalisation of railway maintenance have made it less necessary for staff to stay near the work site and most now travel by road in special staff vans. Thus the number of service coaches in Departmental use nowadays is comparatively small and the vehicles seen in the plates which follow are largely a tribute to a bygone age in our railway history.

Numbering

Coaching stock converted from revenue earning stock for Departmental use prior to 1966 was given appropriate regional letters - DE, DM, DS and DW or the all-regional code DB. The latter was used for later conversions. The numbers added after these letters came from the appropriate series: 320XXX for DE, 395XXX for DM, 70XXX for DS and 150XXX for DW whilst the all - region post 1966 DB coded vehicles had numbers in the 97XXXX series.

In 1973 the TOPS system was introduced and Departmental vehicles were given a prefix letter to indicate which department owned them as listed below.

No prefix - Regional Civil Engineer, (RCE).
A - Regional Mechanical & Electrical Engineer, (RM &EE).
C - BREL.
K - Regional Signal & Telecommunications Engineer.
L - Regional Mechanical & Electrical Engineer Electrical Construction.
R - Research Department.
T - Traffic Department.
X - Stores Department.
Z - Public Relations and Publicity Department.

These owning departments ceased to exist from 6th April, 1992 but the prefixes continued in use.

Note: In 1984 management reorganisation by BR resulted in 'Chief' disappearing and so CCE and CM & EE were replaced by RCE and RM & EE.

CARKIND

This is a three character code normally painted near the vehicle number which show its use. The first letter of the codes in use in this book is generally Q indicating a former coaching stock vehicle, whilst the second letter denotes the use:-

QP	Staff, Dormitory.
QQ	Tool Van.
QR	Stores, Materials.
QS	Miscellaneous Operating Vehicle–Runner, Steam Heating Boiler Van, Loading Ramp.
QT	Brake Van.
QV	General Equipment Carrier.
QW	Self Propelled Unit Saloon.
QX	Specialist Equipment Vehicle–Instruction, Inspection, Mobile Office, Test Coach etc.
QY	Electrification Equipment Vehicle.

The third letter indicates the power brake fitted, the codes being:-

A	Air Brake.
B	Air Brake, through vacuum pipe.
O	No power brake.
Q	No power brake, through air pipe.
V	Vacuum brake.
W	Vacuum brake, through air pipe.
X	Dual brake (air & vacuum).

Liveries

Many coaches merely had their numbers changed when they entered departmental service - unless they were the subjects of major alterations when they were out-shopped in the livery of the time.

From 1948 - 1960	Black all over with yellow lettering.
1960 - 1963	Gulf red body with black underframes and bogies, white lettering.
1963 -	Olive drab (green) with black underframes and bogies, grey roof and

white lettering giving way to grey and yellow and then specialised liveries. Olive green weathered to give different colours which were further affected by the light conditions under which colour photographs were taken. Hence we have olive grey, olive brown etc. The ends of vehicles were variously painted yellow and also given 'wasp' stripes to make them more visible.

Abbreviations used.

Railways regions etc:-

BR	British Railways	BREL	British Railways Engineering Ltd.
CR	Caledonian Railway	ER	Eastern Region of BR
GCR	Great Central Railway	GER	Great Eastern Railway
GNSR	Great North of Scotland Railway	GSWR	Glasgow & South Western Railway
GWR	Great Western Railway	LMR	London Midland Region of BR
LMS	London Midland & Scottish Railway	LNER	London North Eastern Railway
LNWR	London & North Western Railway	LSWR	London & South Western Railway
LYR	Lancashire & Yorkshire Railway	MSLR	Manchester, Sheffield & Lincolnshire Rly.
MR	Midland Railway	NBR	North British Railway
NSR	North Staffordshire Railway	ScR	Scottish Region of BR
SECR	South Eastern & Chatham Railway	SR	Southern Railway or Southern Region of BR
WCJS	West Coast Joint Stock	WR	Western Region of BR

Types of vehicles in this book

Up to three letters are generally used in the abbreviations for a coach.
The class is indicated by:-
F - First, S - Second, T - Third and C - Composite, (F+S or F+T). If this letter is followed by K the vehicle has a side corridor and toilet(s) . Gangwayed open coaches use the letter O instead of K. Non-gangwayed coaches with toilets are indicated by L. Open second or third class coaches are TSO and TTO on the LNER and BR with two seats on each side of the gangway, (TO on the LMS) whilst for a two plus one seating arrangement, they are SO or TO, (RTO on the LMS where they were third class dining cars). On BR, the open dining cars without kitchens were RFO, RSO and RUO, where the U indicates unclassified seating.
Van abbreviations are:-

B	Bogie Full Brake	BG	Bogie Full Brake (gangwayed)
BY	Full Brake (4 wheel)	CCT	Covered Carriage Truck (4 wheel)
GUV	General Utility Van (bogie vehicle)	PLV	Passenger Luggage Van
PMV	Parcels and Miscellaneous Van (also applied to PLV on BR).		

General Abbreviations

ARP - Air Raid Protection
CCE - Chief Civil Engineer
CM & EE - Chief Mechanical & Electrical Engineer
CWMD - Carriage & Wagon Maintenance Dept.
DMU - Diesel Multiple Unit
EML - Euston, Manchester and Liverpool Elect.
GSU - Glasgow Suburban Electrification
MPD - Motive Power Depot
PAD - Pre-assembly Depot
RCE - Regional Civil Engineer
S&T - Signal and Telecommunications
TOPS - Total Operations Processing System
WRD - Wagon Repair Depot
CARMD - Carriage Maintenance Depot
CE - Civil Engineer
CW - Carriage and Wagon Department
DC - Dean/Churchward
ECD - Electrification Construction Depot
EMU - Electrical Multiple Unit
HST - High Speed Train
OHLM - Overhead Line Maintenance
PSS - Power Supply Section
RM & EE - Regional Mech. & Elect. Engineer
TMD - Traction Maintenance Depot
T&RSMD - Traction & Rolling Stock Main. Depot

Plate 1. DM395490 is an 'CELL TRUCK' used to carry old and new accumulators between locomotive/carriage depots and Wolverton Works where these items were serviced. It was converted in 1958 from LMS CCT M35060M which was built to diagram 1871, lot No.123 at Wolverton Works in 1926. Apart form the sealing of the end doors and removal of the bottom end flaps, it is externally as built but note the hole in the side above the lefthand wheel which was to allow for the escape of acid spilled in transit - the floor itself was lined with lead. Scrapped in 1987, it was photographed ex-works at BR Wolverton on 16th November, 1958 in carmine livery. F.W.S.

Plate 2. DW314 is an ex-GWR CCT code named 'Python', built originally to diagram P14 at Swindon Works. It has been converted for use as a Cell Truck for carrying accumulators to and from the works and outstations servicing coaches etc. Lower parts of the body have been plated and the DC brake levers have been replaced by conventional short ones to the central V-hanger. Photographed at Moat Lane on 16th October, 1963. N.

Plate 3. E972022 carries the white lettering 'BR/VACUUM DUST EXTRACTOR/THORNTON DIST.' above its number. It was built by the NBR at its Cowlairs Works in 1896 as a six wheel BS and measures 33ft 6ins in length and has a tare weight of 14.5 tons. Some doors have been taken out of use, but it is little altered otherwise. Note the long early style spindle buffers which are attached to wooden headstocks. Livery is dirty weathered black. Photographed at Stirling Station in August 1956. P.C.

Plate 4. DS70000Es is a compressor unit which was converted in 1959 from the underframe of ex-LSWR third coach No.S353S. It was photographed at Stewarts Lane in January 1976 but in 1981 it was further converted into the CCE 'Britannia' Rail Loading Vehicle based at Sandown on the Isle of Wight. Its CARKIND was QXV (not carried here) and the livery was yellow. D.L.

Plate 5. DE320664 - in use by the Signal Engineer's Department, York as a compressor carrier, is a former GNSR/LNER six-wheeled passenger brakevan which has unusually small, non-standard lettering, one pair of disc and two pairs of spoked wheels and clasp brakes on the outer wheels only. There are footboards at the lefthand end where the doors are in use and it was photographed at York on 10th October, 1959. R.C.

Plate 6. 960860 is an ex-GER six wheel, five compartment brake coach still much as built but with only five roof ventilators and two doors on this side out of use, (no handles). It has an arc roof, long spindle buffers and Mansell disc wheels. Faint lettering below the ducket reads:- 'ENG DEPT/PAINT VAN'. Photographed at Watton, (GER) in the 1950's. R.C.

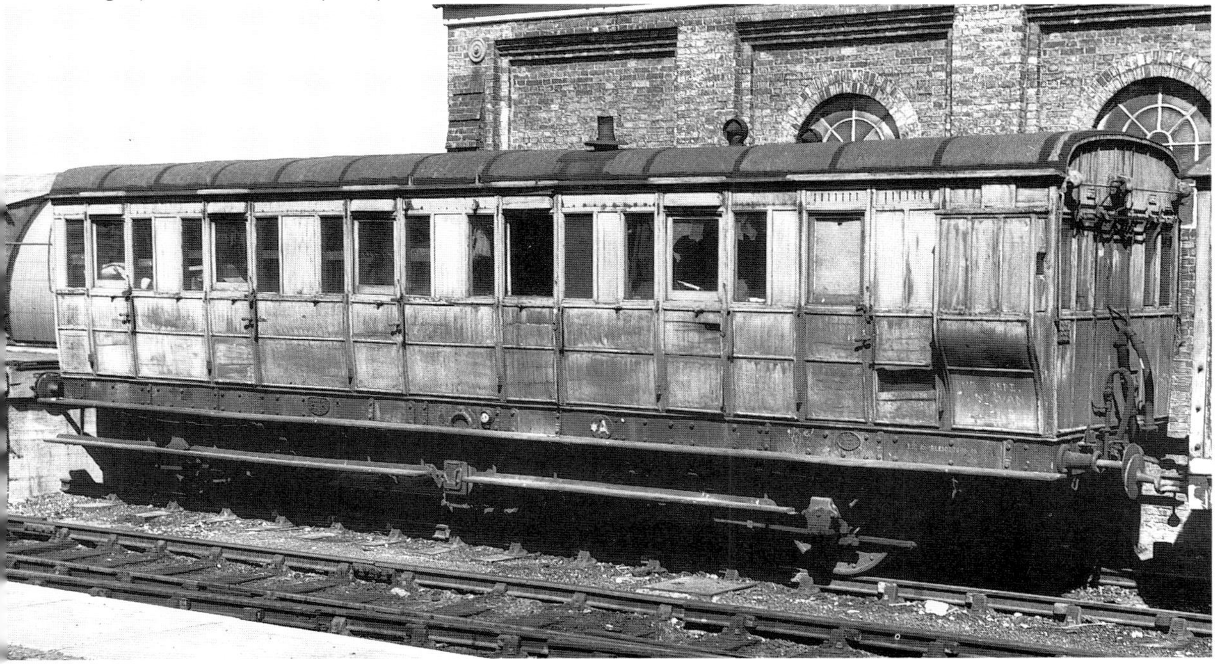

Plate 7. DE320010 is a vintage ex-NBR six wheeled passenger brakevan which has planked body repairs and reduced length footboards. Note the slender spokes of the wheels, long spring J-hanger suspension, wooden headstocks with long spindle buffers and vacuum clasp brakes on the outer axles only. In weathered wood finish it has white lettering on a black patch which reads:- 'BRITISH RAILWAYS/SCOTTISH REGION/DISTRICT ENG DEP/EDINBURGH DIS/PAINTERS PLANT'. Photographed at Edinburgh Waverley Station in April 1952 P.C.

Plate 8. DM395088 is an ex-LNWR CCT built to diagram D.444, which like a lot of this type, has been taken into the Engineering Department as a Stores van. Note the end doors and high arched roof which enabled it to carry tall loads. It is as built with double sliding doors, typical LNWR long springs, axleboxes, handbrake levers and buffers. Clasp brakes on the outer wheels only. Photographed at Chester in 1967. H.F.W./ R.C.

Plate 9. ADM40279 TOPS coded ZRV,(some were later coded QRV), this vehicle is a Stores van working out of Horwich. Lettering at the left reads:- 'M & Em' 'ZRV' 'STORES-M-HORWICH'. It was built as an Insulfish-X to diagram D.2115 with an underframe to passenger traffic standards but also having a handbrake. The body was of double skinned wooden panels which gave these vans longevity and thus many went into Engineering Department service. Note the clasp vacuum brakes on all wheels, robust coach-type buffers and long spring J-hanger suspension with open front axleboxes. Very much as built but in olive green with white lettering, it was photographed at Horwich on 27th October, 1979. R.SL.

Plate 10. DE321107 was a 1966 conversion into a CCE Staff & Dormitory coach for a Tunnel Gauging vehicle which was based at Leeman Road, York where it was photographed on 20th March, 1978. CARKIND QPV, but not carried here, it was converted from BR MkI horsebox E96305 which was built to diagram 751, lot No.30146 at BR Earlestown Works in 1957. Windows have been put in the ends but the side window pattern is unaltered except for the addition of two in the heavily modified livestock compartment. The underframe is basically unaltered. Livery is non-standard green body with yellow ends and doors. D.L.

Plate 11. DW150033 was built as a GWR Passenger Fruit Van, No.2348, which was used to convey mushrooms and soft fruit. It was converted in December 1955 to a BTU Packing Van for use at Banbury, (MP/BANBURY/PACKING/VAN). It is as built except for the loss of the single central pot oil lamp, (which had a conspicuous housing on the roof), and it runs on ordinary disc wheels as opposed to the original Mansell type. Note the long spindle buffers, tall vacuum brake pipe standards and the long springs. The DC brake lever is on the right. Photographed at Banbury MPD on 21st July, 1963, it was condemned in February 1971 and disposed of for scrap from Vauxhall C&W in April 1971. J.M/R.C.

Plate 12. ADW150142. No.2 Fuel Tank wagon consists of two identical cylindrical tanks with filling and discharge gear mounted on the chassis which came from GWR six wheeled milk tank wagon W2501W, which was built to diagram O.37 in 1932. The conversion on lot No.341 in 1957 was originally for use at Ranelagh Bridge and the partially obliterated white lettering on the black body reads:- 'TO WORK BETWEEN OLD OAK COMMON RUNNING & MAINTENANCE DEPOT AND RANELAGH BRIDGE TURNTABLE SIDINGS'. It was photographed at Oxford MPD on 4th August, 1977. G.G.

Plate 13. ADW150142. This plate shows the sealed ends of the fuel tanks when the wagon was based at Reading DMU depot on 2nd July, 1989. S.J.

Plate 14. ADW150142. The discharge gear is seen in this plate taken at Oxford MPD on 16th September,1979. B.D.

Plate 15. ADW150322 is an ex-GW Siphon G, No. 1199 which was built at Swindon Works in December 1928 to diagram O.22, lot No. 1896. It was condemned in January 1962 but soon afterwards conversion took place to a Steam Heating Instruction van. The corridor connection was removed and all except three of the top sections of louvred ventilators were plated over whilst new windows were fitted. The lower parts of the body are much as built except for the plated doors. It still runs on 9ft wheelbase American bogies and was photographed at Oxford MPD on 4th August, 1977. G.G.

Plate 16. ADW150322. This view shows the opposite end of the vehicle where the corridor connection has not been entirely removed. Sitting sadly at Swindon Works with some windows broken, it is now condemned and awaits its fate when photographed on 14th October, 1979. B.D.

Plate 17. ADS70188 is a train heating boiler van which was built in 1962 on the 12ft wheelbase chassis of a mineral wagon. The latter was built to diagram 1386, order A.810 at Ashford Works in 1934 and carried the number S40398S. It has a 'Tare 21t 0cwt/Max speed 45mph' and is 'Not to be loose shunted'. Note the water and fuel gauges at both ends of the body and the added footsteps below the access door. TOPS coded ZSP, livery is faded Rail Blue with white lettering and it was photographed at Swindon Works on 14th October, 1979. By 1990 it stood condemned at Millerhill Yard awaiting disposal to MC Metal Processing, Glasgow. B.D.

Plate 18. ADS70185 is a similar vehicle to ADS70188 and was also built on a mineral wagon chassis - in this case S40224S of the same diagram but order A.678 at Ashford Works in 1931. Carrying the same livery as the vehicle in *plate 17,* here we see the other side compared with the photograph above. Also constructed in 1962, it was originally assigned to Malago Vale, Bristol as a train heating boiler van but it eventually moved to Immingham C&W where it was photographed on 17th September, 1983. R.SL.

Plate 19. TDE321100 carried the CARKIND QXV and was converted to a steam heating boiler van in 1966 from BR MkI horsebox E96327, (built to diagram 751, lot No. 30146 at BR Earlestown Works in 1957). Externally it is as built except for the livestock compartment which has been modified as in *plate 10*. Latterly operated as a Train Heating Boiler Instruction Car it was transferred to the internal user fleet in April 1981. Livery is pale purplish pink with white lettering. Used at Doncaster Works and later at Doncaster Carr - where it was photographed on 1st September, 1991 - it was sold to the Keighley & Worth Valley Railway in May 1992. S.J.

Plate 20. ADM395949 - branded 'Heater Van', is a steam heating boiler van which was converted from an ex-LMS Period I TO of diagram 1745. A considerable number of these coaches were so converted at the change over from steam to diesel traction. The end corridor connections have been removed, the sides have had door and window alterations and the interior was considerably modified for the new use. Photographed at Old Oak Common on 15th October, 1979. B.D.

Plate 21. DS14S is one of the few SR BY vans which were put into Departmental service. It is also of interest being one of a small batch built to diagram 3094, order No. 1091, which had a full-width safe which only had external doors on each side, (the door can be seen to the right of the guard's door). They also had large white numbers painted on red backgrounds on the top left and right of the bodysides - the body colour is Southern green. Unaltered in design, it is seen in 'ENPARTS', (white letters), service at Swindon in Spring 1969. D.L.

Plate 22. DW198 is an ex-GW 10T Mink D express goods van seen here as built except for the steps under the doors. Note the DC brake lever, spindle buffers, J-hanger spring suspension and the central vertical bracing for the flimsy tie rod. With a wheelbase of 18ft 0ins and a tare weight of 9 tons 14cwts, it is seen in use for the carriage of engineering parts for servicing locomotives - 'ENPARTS'. Photographed at South Brent on 2nd June, 1951. N.

Plate 23. ADB975785. CARKIND QRV, this is a BR-built Siphon G of GW diagram O.62 lot No.1721. Outshopped in October 1950 as W1310W from Swindon Works, it was withdrawn in August 1978 and was converted to an 'ENPARTS' van. The lower louvres at each end have been plated over and the gangway connections have been removed but it is little altered otherwise and runs on 9ft wheelbase heavy steel bogies. Livery is olive green with white lettering - the lettering quoting the working limitations having been applied freehand. Condemned in March 1985, it was photographed at Penzance depot on 13th July, 1983. It subsequently became Internal User 061047 and was based at Plymouth (Tavistock Junction). S.J.

Plate 24. ADB975613. CARKIND QRX, this coach was formerly BR MkI BG S80918 which was built by Pressed Steel in 1957 to diagram 711, lot No.30162. It was converted to TLV S68202 and then to a stores van on lot No. 3917 in 1976. Here working as an HST ENPARTS VAN, it shows no major external changes after all its conversions and was photographed at Didcot on 1st October, 1979. By 1985 it was coded QQX and was a Breakdown Train Unit tool van at Old Oak Common. B.D.

Plate 25. ADB975041 was built as BSO No.W9266 to BR diagram 183, lot No.30170 at BR Doncaster Works in 1956. It was converted to an ENPARTS vehicle after withdrawal from revenue service in 1970. Unaltered externally, it operated between Swindon Works and various WR depots and is seen here at Old Oak Common on 26th October, 1979. B.D.

Plate 26. ADB975039. Here we see the other side of BSO W9264, with the same build details as the vehicle seen in *plate 25*. It had the same service use as ADB975041 but was still carrying its original maroon livery here. Both were condemned by 1982. Between March 1982 and May 1983 the ENPARTS service became part of Speedlink and its vehicles were largely disposed of or re-used. Photographed at Landore depot on 10th September, 1974. B.D.

Plate 27. ADW150199 is based on the underframe of horsebox W263 which was built at Swindon Works in 1928 to diagram N13, lot No.1397. Conversion took place in May 1959 to a CM&EE wagon to transport diesel engine parts, (DENPARTS), between Swindon Works and other depots and workshops. It was condemned at Swindon in December 1981 and scrapped at Woodhams, Barry in February 1982. The unaltered underframe has had three wooden crates fixed to the floor which are in weathered wood livery with white lettering and some black patches. Photographed at Swindon Works on 14th October, 1979.B.D.

Plate 28. ADB975562 is an ex-LMS Period III BG which was built to diagram 2007, lot No.1096 at Wolverton Works in 1938 and carried the number M30976 originally. It stands out of use at Ruabon on 9th August 1989 - sadly with most of its windows broken - not having seen any real service since 1982. When converted from revenue stock it became a Stores van, CARKIND QRV, and worked between Crewe Works and Glasgow Salkeld Street carrying diesel engine parts - DENPARTS. Happily it was destined for preservation on the Llangollen Railway. A.T.

Plate 29. DB975503. TOPS coded YKA and code-named 'Manta' this is a long welded rail carrier and it is included here because it is one of eight which were converted on lot No.3717 in 1970 from Southern Electric 6-PAN TFK's of diagram 2506 (in this case S12263S) built in 1935. The 61ft 9ins underframe has been used and bolsters have been mounted on the floor to retain the rail. All these vehicles were based at Redbridge RCE depot, Southampton. By 1990 all eight were TOPS coded YNA and were S&T general materials carriers based at Selhurst T&RSMD. All had been withdrawn by November 1993. Photographed at Three Bridges on 8th May, 1983. D.L.

Plate 30. DB975526 is one of 26 vehicles with the same conversion details as the 'Manta' seen in the above plate, except these vehicles originated from a variety of 4 car electric unit coaches. This one came from 4-COR TSK S10064S. These conversions were code named 'Marlin' and like 'Manta' were 14 ton bogie rail wagons but the length here was 66ft 3ins. Note the rail loading ramp (lettered 'Speedlink (UK) Ltd/Parkeston Quay'), original buffers and bogies. By 1993 twelve were still extant. Photographed at Three Bridges on 8th May, 1983. D.L.

Plate 31. DB977200 - built as a BR MkI CCT to diagram 816, lot No.30549 at BR Earlestown Works in 1959, it originally carried No.W94444. It was converted in 1984 at Cardiff Cathays for use as a WR RCE tool van with Rail Profiler Machine No.DX79105 – by 1990 it was recorded with DX79104. Livery is olive green with white lettering but the lefthand door is in Rail Blue (damage replacement), and the underframe is black. Photographed at Carlisle Currock on 7th December, 1990. A.T.

Plate 32. ADB977867. CARKIND QXA, this coach is in use as a training vehicle and has been little altered, externally. It was constructed as MkIIe TSO No.W5841 at BR Derby Works in 1972 to diagram 108, lot No.30844. Conversion took place at Doncaster Works in 1993 and it was based at Preston Station. Retaining its original livery with added black lettering, it was photographed at Derby Station on 10th August, 1984. A.T.

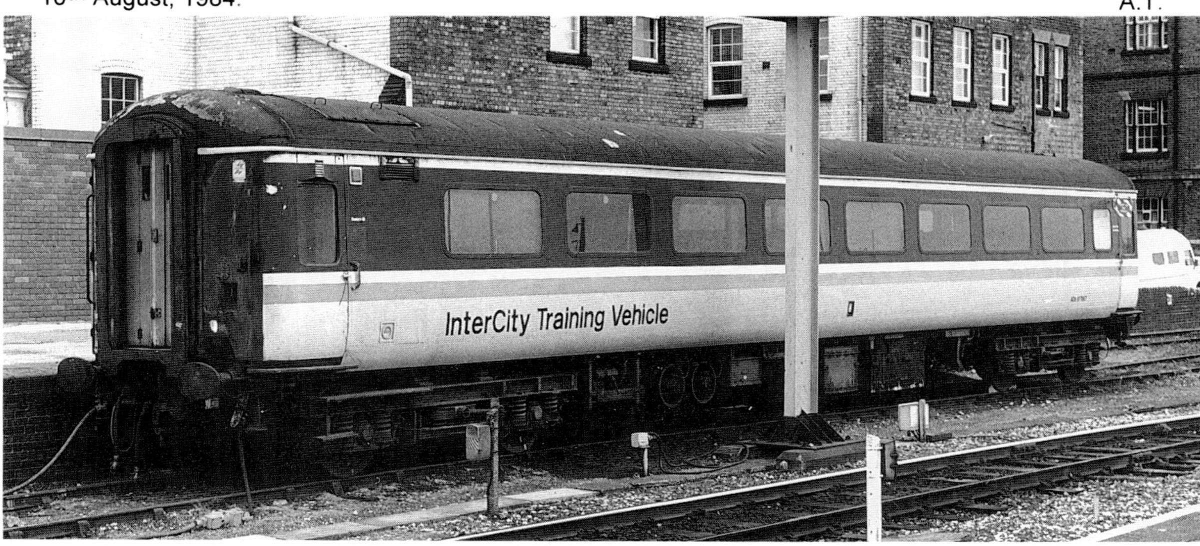

Plate 33. ADB975665 is a 'WESTERN REGION HST BARRIER VEHICLE' used in connection with the movement of HST coaching stock by non-HST motive power or vice versa. It was converted in 1978 from BR MkIIA BFK No. W14071 which was built to diagram 163, lot No.30775 at BR Derby Works in 1968. Externally it is very much as built including B4 bogies and it carries Rail Blue livery with white lettering. CARKIND QSA but not carried here, it was photographed at Didcot on 1st October, 1979. B.D.

Plate 34. ADB975665 is seen here on 21st September, 1991 at Doncaster Decoy Yard from where it was sold to Maize Metals of Wednesbury who scrapped it on site. In the same livery as in the plate above, it has had its windows painted over and runs on BR MkI bogies. Note how the corridor connection has been boarded up internally. A.T.

Plate 35. ZDW150353 originated as BR MkI BSO No.W1012 which was built to diagram 56, lot No.30014 at the BR York C & W Works in 1951. It was converted in 1963 to a Publicity Department Cinema Coach based at BR Wolverton but for use all over the BR system. The coach is little altered externally here and its livery is Rail Blue and grey with white lettering. Photographed at Salisbury on 9th July, 1980. R.SL.

Plate 36. ZDW 150353. This view of the above coach shows the opposite side with the toilet window nearest the camera. Here the side has been modified by removal and plating over of three of the central windows and it sports faded inter-City livery. Photographed at BR Wolverton Works on 1st October, 1988. A.T.

Plate 37. M395017. 'CINEMA COACH/No.2' started life as a twelve wheeled 65ft 6ins LNWR Sleeping car which was built at Wolverton Works in 1907. This arc roofed design to diagram D16 was the LNWR's final type of this length of Sleeping car and only six were built. All were scrapped except this one which has undergone radical changes for its new role. Doors and windows have been plated over but sadly few photographs exist of it and the later pattern of six wheel bogies which were fitted cannot be seen here. Freshly out-shopped in black livery, it was photographed at Wimbledon in 1956. P.C.

Plate 38. ZDB975403. CARKIND QXX, is also a Cinema Coach but the fully windowed side of the original coach has been almost panelled out here. It was a BR MkI TSO which was built to diagram 93, lot No.30243 at BR York C&W Works in 1956 and conversion was on lot No.3865 in 1975. Operating on all regions, it was originally based at BR Wolverton and was photographed at Longsight Manchester on 27th February, 1982 when its livery was Rail Blue with white lettering and black underframe. SJ.

Plate 39. DE951550 is another of those early coaches given an extra lease of life by various MPD's as stores, staff accommodation, tool vans etc. often in breakdown trains. This one was allocated to 'MOTIVE POWER DEPT./TRAFFORD PARK' where it was photographed on 12th March, 1955. Built as a MSLR three compartment passenger brakevan, it is in very good condition with fine original wooden panelling, tall vacuum brake standard, spindle buffers and Mansell disc wheels - note how the light has caught the inside detail of the righthand wheel. F.W.S.

Plate 40. DE320608 is an ex-GCR/LNER saloon coach with typical matchboard sides, windows in the end and coach roofboard holders. Lettered 'MPE GRANTHAM', 'STAFF RIDING VAN', it was a rather special vehicle for such a use and hasn't undergone any major external changes in its conversion from revenue to departmental stock. Note the long spindle buffers with fluted housings, slender truss rods and 10ft 6ins wheelbase bogies. Such vans were used in breakdown trains and they will receive more coverage in a future volume of this series. Photographed at Grantham on 9th August, 1959. R.C.

Plate 41. DE960822 is an ex-Pullman car which was built at the Clayton Wagon Works in 1922. A rather luxurious vehicle for use as an Engineering Department Staff Coach which has not undergone much radical alteration externally apart from the loss of corridor connections which have been boarded over as have the doors and small windows in the ends. The roof details are unaltered - note the brackets to hold the roofboards - and it runs on the original American style six wheel bogies which have LNER open front axleboxes. Livery is black and it was photographed at Cambridge MPD in the late 1950's. R.C.

Plate 42. 970201 is branded 'BR/SLEEPING & MESSING VAN/EsC/INVERNESS' in white lettering on the black body. Like the previous vehicle, it is an ex-Pullman Car of similar vintage and it has undergone similar external alterations. The bogies are original and carry the date 1920. Photographed at Inverness on 31st May, 1956. F.W.S.

Plate 43. TDE321108 - probably the last Medical Coach on BR when it was photographed at Whitemoor on 27th June, 1980. Finished in olive green, it has had the end gangway connections removed and boarded over and a handbrake lever has been fitted. Note the typical Gresley design features including clipped buffers, panelled teak body, bogies and roof profile. It was built as TTO No.56856 to diagram 186 by Metro-Cammell in 1938, but its appearance here is now different. Purchased by RESCO Railways in 1981, sold to the LNER Carriage Association in 1988, it was restored at Starbeck and is preserved on the North Yorkshire Moors Railway. R.SL.

Plate 44. DE320202 is an ex-GCR/LNER third class saloon coach which was one of three, (No.s 1315-1317), built at Gorton works in 1900. Externally it is as built except that the entrance door to the right of the number has been taken out of use and there is no gangway connection on the lefthand end. It is in use as an 'INSTRUCTION CAR' based at Doncaster C&W. The saloon section with longitudinal side seats extended the length of the coach except for toilets either side (small window over number) and vestibules at each end, the lefthand one having double outward opening doors. It runs on 8ft wheelbase bogies set at 33ft 6ins centres and note the slender truss rods and the obtrusive central handbrake wheel. Photographed at Doncaster Works Yard on 30th May, 1959. R.C.

Plate 45. No.7500 is an ex-MR arc roofed all third six wheel coach built at Derby Works circa 1886. In use by the Motive Power Department, it still has the original end panelling but little is left of that on the sides. Its dated underframe is still largely unaltered with wooden headstocks carrying quaint spindle buffers, long spring J-hanger suspension and split spoked wheels. Livery is weathered wood with white lettering and it was photographed at Peterborough MPD on 25th March, 1950. R.C.

Plate 46. DM297283 This superbly maintained coach is a conversion from a CR six compartment brake where several windows have been discreetly boarded over and only one door is in use on this side. It still retains its original bogies and buffers. Built at St. Rollox Works in 1899, it was photographed at Wick in September 1961. N.

Plate 47. DM297291 is an ex-GSWR/LMS six wheel passenger brake van built in 1904 and in use here as an Engineer's Staff Coach based at Inverness. It has replacement sections to its panelled sides and planked ends. The wooden headstocks carry long spindle buffers and three link couplings. Note the different length of spring hangers on outer and centre axles, two pairs of Mansell disc wheels and end windows which unusually open externally by sliding downwards. Photographed at Forres on 14th June, 1960. R.C.

Plate 48. M12328 is an ex-NSR picnic saloon which was originally fitted with seats along the sides. Some doors are out of use and windows have been boarded over but it is largely in revenue earning form and is pictured here in faded crimson livery when in use as a riding/mess van at Stourton MPD on 8th July, 1956. F.W.S.

Plate 49. DM293561. The LYR would not appear, from the map, to be a large railway but it served several densely populated areas and therefore required large quantities of rolling stock. Here we see a BTK built at Newton Heath Works in 1898. Apart from repairs to bodywork and doors taken out of use, it is much as built - note the large ducket and original bogies and buffers. Its peeling paint is light brown and the white lettering reads :- 'PERMANENT WAY DEPt', 'BARROW'. Photographed at Barrow-in-Furness Shed Yard on 6th May, 1954. F.W.S.

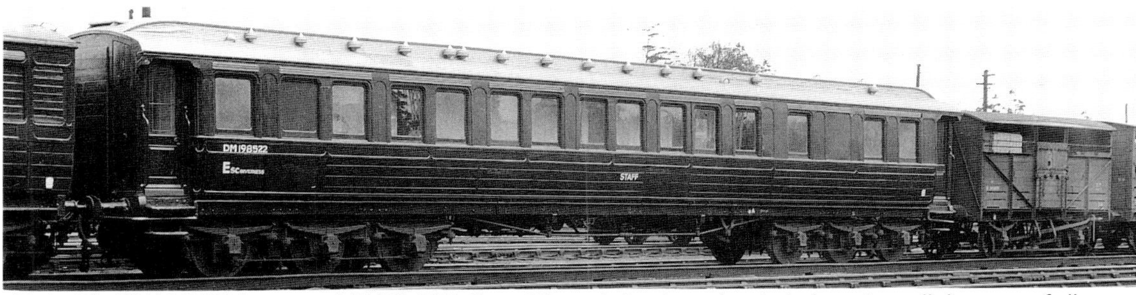

Plate 50. DM198522 is an ex-LNWR 65ft 6ins, 30 seat twelve wheeled clerestory dining car of diagram 35A/39 which was built at Wolverton Works in 1903/4. As one of the last three survivors of 20, it was converted to an ARP lecture car by the LMS in 1938 and subsequently became a Staff Coach as seen here. Corridor connections have been removed and much of its fine panelled body has had to be replaced but the underframe details and clerestory features are very much as built and its livery is black. Photographed at Inverness on 15th May, 1958. F.W.S.

Plate 51. DM198522. Taken at the same time as the previous plate, this view shows the other side of the same vehicle with its different window pattern. Three windows have been taken out of use on this side and two on the other side. F.W.S.

Plate 52. M198629 was built at the MR works at Derby in 1906 as a 65ft Dining Car. The windows on the sides and the clerestory have been plated over, only two doors remaining (the doors themselves are different from the originals). By comparison, little alteration has taken place below the solebars and it was photographed in newly painted black livery at Inverness in August 1956 in use as a Tool Van. P.C.

Plate 53. DM198931 is an ex-WCJS/LNWR 65ft 6ins twelve wheeled clerestory sleeping saloon - one of five which were converted into sleeping/mess vans for the SR in 1949. One was allocated to Seaton, (Devon) and is seen here in 1954/55 on its own stretch of track where it remained until scrapped in 1959. The ornate ends have had the gangway connection removed. The corridor side is seen here - minus one window, (there were ten equally spaced). J.M./R.C.

Plate 54. DM198932 is one of the other four sleeping/mess vans which, like the one in *plate 53*, was converted from a 65ft 6ins twelve wheeled clerestory sleeping saloon in 1949. This one is lettered 'MP WADEBRIDGE' which is where it was photographed in 1958. It is very much as built externally except for the usual changes to the ends and it has the original set of windows along the corridor side which is seen here. R.S.

Plate 55. DM354879 - a former MR twelve wheeled 65ft first class clerestory sleeping coach built at Derby to diagram D451 in 1905. Although the gangway connection has been removed it hasn't been plated over externally. The sides are very much as built with the same window pattern but sadly most of the lower panelling has had to be replaced by planking. The underframe and fittings are as built and its livery was black when photographed at Aberdeen Station on 14th May, 1958. F.W.S.

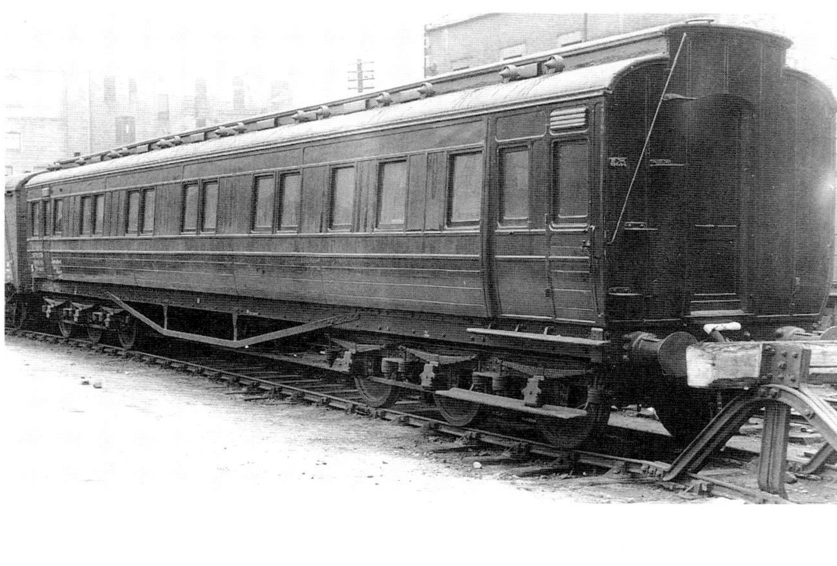

Plate 56. TDM395092 was built by the MR at Derby works in 1913 as a 31ft six wheeled passenger brakevan to Bain's diagram 530A lot No.788, (original number 184,1933 No.34147). It was withdrawn in December 1951 and converted in 1954 for service in a weedkilling train. Subsequently it became an RCE Staff and Dormitory van based at Guide Bridge, for use with crane DRT81264. By 1988 it was out of use and by 1990 it stood condemned at Carlisle Kingmoor Yard. Livery is olive green body with white stencilled lettering and black underframe. Photographed at Carlisle Kingmoor on 2nd June, 1988. A.T.

Plate 57. DM395867 has undergone some radical changes from when it ran as M5757M - an LMS Period III BTK, which was built at Wolverton Works in 1935 to diagram 1905, lot No.859. It was converted in 1964 to work as a Staff and Dormitory coach with the twin jib crane, DRC78236 in which role it was photographed at Northampton Bridge Street on 27th October, 1989. It is finished in smart Rail Blue livery with white lettering and red buffers and headstocks. Note the footsteps picked out in white. S.J.

Plate 58. ADB975876 is lettered "CM & EE(LMR) Training Coach" and the recessed door with special handrails shows its origins as Pullman kitchen first "HAWK", (later E314E), built in 1960 by Metro-Cammell. It still looks smart in Rail Blue and grey livery with black running gear. (The Commonwealth bogie springs are red and the roller bearing covers are yellow). Converted on lot No.3974 of 1979, it was used at Oxley Carriage Sidings but by 28th May, 1988 it was photographed at Leicester awaiting its fate. Photographs appeared in 'Rail' magazine No.91 (March 1989) and No.202 (June 1993). A.T.

Plate 59. DB975807 was built as SR BY passenger guard's van No.S683S to diagram 3092 on order A.974 - a joint Ashford/Eastleigh project of March 1938. It was converted at Stewarts Lane WRD in July 1979 on lot. No.3963 to replace DS22, *(see Frontispiece),* as an RCE 'GAUGING VAN' (under these words is painted 'HEDGEHOG'!!). Major body surgery has taken place to allow for the fitting of the gauging frame but what remains is much as built except for the window apertures cut into the lone end. Body colour is very pale cream with a light grey inner end, black lettering and the gauging frame etc. is picked out in white. The other side is similar and is pictured in 'Rail' magazine No. 145, April 1994. Photographed at Woking on 4th May, 1991. A.T.

Plate 60. DS47452 is not a coach conversion but it is included here to show how very basic mess and tool van conversions could be. Originally built as a standard SR 12T goods van, it was converted in 1958. Externally, two windows were fitted in each side and the even planked body is black with yellow lettering - the lower part reads 'NOT TO BE/LOOSE SHUNTED'. Based at Faversham, it was photographed at Strood in Spring 1968. D.L.

Plate 61. Ds914S is an ex-SE&CR six wheeled passenger guard's which was built in 1901 and subsequently carried SR No.569. It was one of several vans which were put into Departmental use in 1935-36 as Locomotive Department Mess and Tool Vans. Allocated to Eastleigh, it is much as built apart from the chimney but it looks sadly neglected with its peeling paintwork when photographed at Eastleigh Works circa 1952. H.F.W./R.C.

Plate 62. DB975566 is an ex-SR designed PMV which was built to diagram No.3103 in July 1951. Numbered S1496S then, its external appearance here is little changed. It is one of five such vans converted to electric lighting in 1965/66 and in 1967 it was one of ten air-braked for Central division work. In 1978 it was given its DB number, one of only a handful of such vans to get these numbers. As a tool van it was used at PAD's such as Ashford and Three Bridges before moving onto the WR. Livery is Rail Blue with some black patches and it was photographed at Radyr on 11th July, 1978. R.SL.

Plate 63. ADB975142, CARKIND QPV, is a Staff & Tool van operated by the PSS. It was built as S7125 - a BY van to diagram 3092 on order A.974 at Eastleigh Works in June 1938 (Ashford body, Eastleigh underframes). Repaired in 1944 after enemy damage in the Second World War, it was converted to the use shown in 1971 on lot No.3779. By November 1993 it had been withdrawn and disposed of to Booth Roe Metals, Rotherham. Externally it is very much as built with wide and narrow planks but it lacks chalkboards, roof vents and lookouts and tall vacuum brake standards. The ends have had three windows put in them and the livery is olive green with white lettering. Allocated to Horsham, it was photographed at Fratton on 2nd July, 1989. S.J.

Plate 64. DS70076 was built in December 1930 to SR diagram 3098 as a bogie luggage van No.S2339S using the underframe from LSWR composite coach No.3001 which was built in May 1905. Withdrawn in December 1959, it was converted soon afterwards into a Mess and Tool van as seen here. Livery is olive grey and the main changes are the removal of corridor connections and addition of windows to the righthand pair of doors. It was finally withdrawn in 1981 and was purchased in May 1982 by the Kent & East Sussex Railway for preservation. Photographed at Hoo Junction in Spring 1969. D.L.

Plate 65. ADS70141 - CARKIND QPV, was built as S2462S in August 1931 to SR diagram 3099 on the underframes of LSWR composite No.2620 (built July 1910). Coded GBL by the SR, it was withdrawn in October 1960 and reinstated as a Staff & Tool van in August 1961. In olive drab livery it is very much as built except for removal of end gangway connections. Allocated to 'ME's/POWER SUPPLY SECTION' it was withdrawn in 1980 and sold from Basingstoke to the Bluebell Railway for preservation in June 1981. Photographed at Cosham on 2nd February, 1978. R.SL.

Plate 66. DS70207 was converted from CCT S2040S which was built to diagram 3101 under order A.3702 at BR Ashford Works in 1951. It was converted in 1963 to a BTU Tool van and on further conversion it became a CCE Tunnel Inspection van as seen here in the snow at Huddersfield in February 1979. Its height has been considerably reduced by cutting the arc out of the roof, the wooden panelled sides have been opened out except at the left and a set of side doors have been removed. Four wheeled luggage van ends have also been fitted and it is finished in black livery. After serving from various West Yorkshire bases, it was condemned at Wakefield in 1985. D.L.

Plate 67. DS1719 is an ex-SECR 45ft three compartment 'birdcage' brake third coach which was built as No.2302 to diagram S2780/1 by the Metropolitan Carriage & Wagon Co. in August 1900. It became SR No.3256 in October 1926, (SR diagram145), and was withdrawn in April 1942, when it was transferred to Service Stock and renumbered 1719S, later becoming DS1719 under BR ownership. Always allocated to the S&T Dept. as a 'MESS and SLEEPING VAN', it spent a lot of its time on the WR and when withdrawn circa1968, it was the oldest of its kind in service. Apart from replacement body panelling, it is very much as built but only one compartment door is in use on the side seen here. Livery is black with cream lettering and it was photographed at Hatton (ex-GWR) in August 1966. R.C.

Plate 68. DS70164 CARKIND QPV, was converted in December 1962 for Emergency Control Train use and was a South Lambeth Office coach in 1981 before coming a CCE Tunnel Inspection Staff coach. It was built as Maunsell BTK S36915 to diagram 2105 at Eastleigh Works in 1931. The corridor connections have been removed and only the brake end door is in use. All windows have been fitted with shields and it carried olive green livery when photographed at Hoo Junction on 1st May, 1982. By 1984 it was coded QPW and it was still active in November 1993. D.L.

Plate 69. DS70164. This plate shows the other side of the vehicle above but this time photographed from the brake end. As before, the original pattern of doors and windows is still evident but only the brake end doors are operational. Photographed at Ashford SR on 7th June,1992. R.SL.

Plate 70. ADB975663 is a Tunnel Inspection van which was converted from SR GUV No.S4593S on lot No.3933 in 1978. The GUV was built to diagram 2182 on order A.975 as a joint Ashford/Eastleigh project in 1938. It carries CARKIND QXW here and is finished in yellow with black lettering. The inspection platform has replaced about 60% of the original bodywork but it is little altered below the solebars. It was still in service with Network South East in November 1993. A.T.

Plate 71. TDS70159 was converted from Maunsell BSK No.S3680S which had been built to diagram 2105 at Eastleigh Works in 1929 and withdrawn in November 1962 for Emergency Control Train use. It saw service as a South Lambeth office coach and later became the Staff coach for Tunnel Inspection Vehicle DM721211. Note only one operational door on the side seen here and the added handbrake lever with its solebar – mounted pivot. The interior is open for its office use. By November 1993 it stood condemned at Hoo Junction. Livery is olive green and it was photographed at South Lambeth on 19th April, 1980. R.SL.

Plate 72. DS70245 is an unusual conversion into DS stock from ex-LMS Period III BCK M6869M which was originally built to diagram 2010, lot No.1098 at Wolverton Works in 1938. The coach shows only minor alterations here - removal of corridor connections and addition of handbrake. It was originally converted for use as an office coach at Stewarts Lane but by Winter 1968/69 when it was photographed at Hoo Junction, it was a CCE Staff and Dormitory coach working with Ballast Cleaner No.5. It continued in this role with DR76214 until June 1987 when it stood condemned at New Cross Gate WRD. Livery here is olive green. It was sold to William Smith of Wakefield but was vandalised on site and was therefore scrapped at Woking PAD by Maize Metals of Wednesbury. D.L.

Plate 73. 1831S. Also seen at Seaton standing next to DM198931 *(in plate 53),* was this ex-LSWR restaurant car mounted on 'Ironclad' coach bogies. This coach had seen service in 1943 as an ARP Clean Dressing Van coupled to ex-SECR six wheel guard's van No.7842. Here it has become 'DORMITORY SET No.6' in the service of the 'EXETER DISTRICT ENGINEER' and appears little changed externally although it originally had corridor connections and the vent in the end is a conversion feature. Photographed at Seaton, Devon on 20th August, 1954. H.F.W./R.C.

Plate 74. DS70024 is a CCE Mobile Kitchen Coach which was converted from ex-LSWR 'Ironclad' BTK S3180S which was built at Eastleigh Works to LSWR diagram 24(c) (SR diagram 135). This plate shows the compartment side and note the typical bogies and buffers and the plated over end corridor connections. Based at Hither Green, it is finished in olive green with white lettering. It was condemned in October 1979 and was scrapped by King of Snailwell. Photographed at South Lambeth on 12th March, 1978. R.SL.

Plate 75. W81W is an ex-GWR Dean 4 wheel composite coach which has been converted to a Tool van for the use of 'OSWESTRY ELECTRICIANS'. Two of the four doors are out of use on this side and defective wooden panelling has been plated over. A stovepipe chimney has been fitted but the only other alteration is to the footboards to accommodate the handbrake lever and its bracket. The small lettering reads:- 'MAXIMUM WEIGHT INCLUDING TARE/NOT TO EXCEED 14 TONS/BRAKE LEVER IN CENTRE'. Livery is dirty black and it was photographed at Porthmadog in 1958. P.C.

Plate 76. W14913 in service as a Staff & Tool coach and based at Plymouth was built by the GWR at Swindon Works in the late 1800's as an all 3rd vehicle. There is little left of the original panelling on the body and only two doors are operational. Note original gas lamp tops on the roof and an added stovepipe chimney. The vacuum brake cylinder, DC brake lever, commode handles and spindle buffers are all original. All the staff section windows are open on a hot day in June 1958 when it was photographed at Par. R.S.

Plate 77. DW14192 converted to a Staff Coach for the Engineering Department and based at Shrewsbury was built at Swindon Works as a 46ft clerestory – roofed, non-corridor all third coach. Some doors have been taken out of use and the footboard reduced accordingly. Handbrake levers have been added and it runs on the original Dean bogies. Note the fine end detail including the feed pipes to the gas lamps which can also be traced along the roof of the clerestory. Photographed at Gresford Colliery Sidings on 1st February, 1962. N.

Plate 78. DW14572 is one of a pair of ex-GWR Dean clerestory non-corridor eight compartment third class coaches which were converted at Swindon Works in 1959 to mess vans for ballast trains for the use of the District Engineer at Wolverhampton. This one was delivered to Inspector Tyler at Stourbridge Jct. The nearest compartment was his office and the rest was the mess and cooking area for the men in his charge. This coach retained its gas lighting and the lamp tops can be seen on the roof as well as three stovepipe chimneys. The other coach was delivered to Inspector James at Wolverhampton Low Level. Photographed ex-works in black livery at Dudley on 10th May, 1959. P.C.

Plate 79. DW150243 is an ex-GWR Collett Autotrailer completed at Swindon Works in January 1929 as No.165 to diagram A27, lot No.1394 - one of twelve vehicles Nos. 159 - 170. Measuring 59ft 6ins by 9ft, it has a steel panelled body and runs on Collett heavy duty 7ft wheelbase bogies. In December 1960, it was converted into an Office/Work Study coach for the Commercial Department in which capacity it served until it was condemned in September 1967 and disposed of for scrap from Yeovil Junction in May 1970. Pictured here shortly after conversion, it shows little change externally except for the removal of the 'gong' from the driver's end. R.C.

Plate 80. DW150044 is an ex-GWR Churchward TK No.3925, which was built in 1915 to diagram C.32, lot No.1246 at Swindon Works. It was converted in September 1956 to a CCE Staff & Dormitory coach for Welders of the Mechanical Relaying Gang based at Shrewsbury where it was condemned in July 1972 and scrapped the same year at Bridges of Wednesbury. Here we see the corridor side with a fragmented running board, one large window and its two toplights plated over and a central handbrake lever fitted. The corridor connection has also been removed and plated over. Note the 8ft American bogies, square shank buffers and heavy duty truss bars. Photographed at Abermule on 11th October, 1962. N.

Plate 81. DW150010 is a similar coach to the one seen above, but numbered 3660 and built to diagram C.31, lot No.1294 in 1921. After use as a Ambulance Train coach in the 1914-18 War it was rebuilt to diagram C.31. Here we see the compartment side which is very much as built except for the added handbrake lever. It was converted in January 1956 to a CCE Staff & Dormitory coach and initially went to Caersws where it was photographed on 11th October, 1962. Transferred to the LMR in March 1963, it was condemned at Crewe in September 1967 and scrapped by Loom of Spondon in January 1969. N.

Plate 82. KDW150271 is a Staff and Tool Coach for the S&T Department based at Reading Signal Works. It was converted from Collett SK No.W4556W which was built to diagram C.54 lot No.1352 in 1925. This view shows the compartment side where two windows have been plated over. The bow end has lost its corridor connection but the roof detail is unaltered. Note the overflow pipes for the water tank in the roof. Livery was dark green with white lettering when it was photographed at Newbury on 7th September, 1975. R.SL.

Plate 83. DW150209 like the vehicle above, was converted from an ex-GW Collett 57ft bow ended SK, in this case No.W4872W, which was built to the same diagram but lot No.1372 at Swindon Works in 1926. Conversion to a Staff & Dormitory coach took place on lot No.1030 in 1959. This view shows the corridor side of the coach which is unaltered apart from the new lettering and numbering which is white, an olive green body and the usual added handbrake lever. Based at the S & T works at Reading, it was photographed at Oswestry on 8th July, 1965. By 1983, it was out of service at Shrewsbury Engineer's Yard. Condemned in 1984, it was then preserved on the Severn Valley Railway. N.

Plate 84. DW150338 was a 1962 conversion on lot No.1125 to a Staff & Dormitory coach from ex-GW Collett 57ft flat ended BTK, W5786W which was built to diagram D116, lot No.1490 in 1933 at Swindon Works. Assigned to 'Ew SHREWSBURY' here, but by 1983 it was the support vehicle for the Ballast Cleaner No. DR76206, moving on to DR76219 by 1985 . Photographed at Oswestry on 8th July, 1965, it was out of use at Crewe Gresty Road by 1987 and was one of the last four Collett coaches to survive on BR. This view shows the corridor side which has had one large window removed (to the left of the loading gauge), and a handbrake fitted. The flat ends have also had their corridor connections removed. N.

Plate 85. This is an ex-GW 'Toplight' multibar, three compartment BTK built to diagram D56, (seven lots built between 1914 and 1921). As built externally and internally, several were put into the 'ENGINEERING DEPT./FOR TEMPORARY USE ONLY' (as lettered on the side here). The number of this one was not recorded when it was photographed at Taunton in 1960 finished in 'blood and custard' livery. It is standing on a siding laid with double-headed 86A rail with inside keys and the 1920"s standard GWR buffer stop is made of the same rail. P.C.

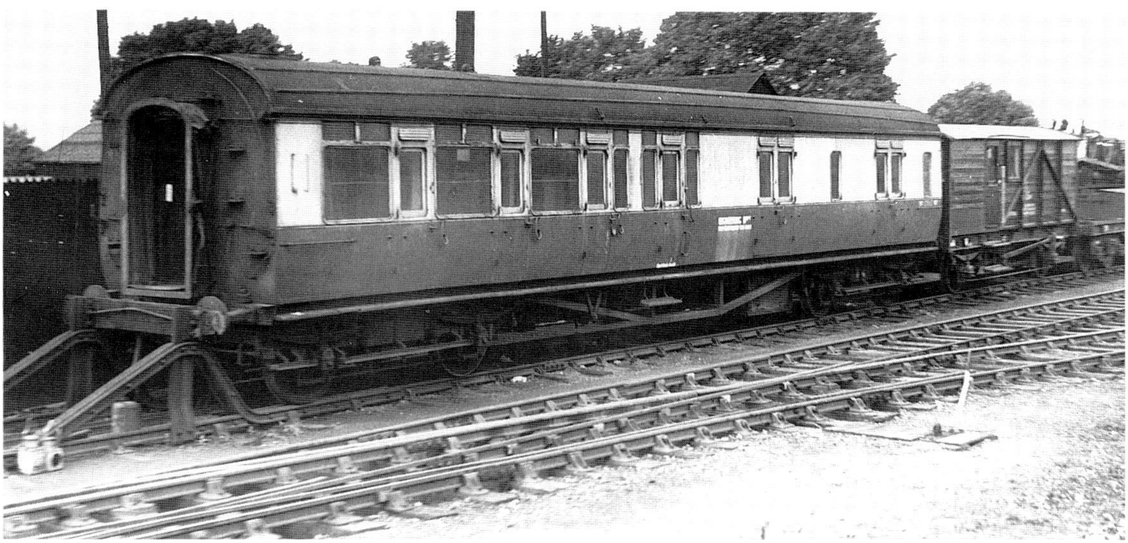

Plate 86. DW150252 is an ex-GW Collett 58ft 4½ins SK (No.4877) which was built to diagram C.54 at Swindon Works in 1927. This view shows the corridor side with three doors out of use, footboards only under operational doors and one door and one and half windows panelled over. Note the central handbrake lever, square shank large headed buffers, 7ft wheelbase bogies and the end steps to access the water tank in the roof. Photographed at Snow Hill Station, Birmingham on 27th July, 1964. R.S.

Plate 87. DW150252. Here we see the compartment side of the vehicle seen in the plate above. It was converted in July 1961 into a CCE Staff & Tool coach for the Wolverhampton District. One door and one window have been plated over and only one door is operational. Lettering at the far end is 'DW150252/Ew WOLVERHAMPTON DISTRICT' and at the near end 'MESS AND TOOL COACH/WOLVERHAMPTON DISTRICT' and 'MAXIMUM WEIGHT INCLUDING TARE/NOT TO EXCEED 30 TONS'. Finished in standard red livery, it was condemned at St. Helens in October 1972 and scrapped at McWilliams, Shettleston in December 1972. Photographed at Hatton in August 1966. R.C.

Plate 88. DW150323 was formerly a GW Siphon G numbered W2798W which was built to diagram 033, lot No.1578 in January 1937 at Swindon Works. Converted for Ambulance Train use in 1939, it was rebuilt to diagram M34 when it returned to revenue service. In Ambulance Train use, all the top louvred vents were plated over except for one in each pair of doors and there were three as windows in one side and two in the other side but the window pattern seen here does not correspond to that assigned in diagram M34, otherwise it is externally, as built. CARKIND QXV, it was converted for Emergency Control Train use and is seen here at Salisbury on 9th July, 1980 in smart maroon livery as an Exhibition Train Generator Van. R.SL.

Plate 89. DW150255 was converted in July 1961 to a CCE Staff & Tool coach for the Shrewsbury District as seen here. Built in 1926 as GWR No.4862, it was a bow-ended Collett TK of diagram C.54, lot No.1372. This view shows the corridor side with the conversion alterations to the regular window pattern near the far end and the other usual changes to ends, running boards etc. Condemned at Derby in February 1970, it was scrapped by Loom of Spondon in April 1970. Photographed at Caersws on 11th October, 1962. N.

Plate 90. DW150046 was converted in June 1956 from ex-Collett Saloon coach No.9372 which itself was a conversion from an ex-Ambulance Train car of earlier GW build. It has two saloons seating 40, one ordinary compartment, a guard's compartment and two toilets. A reminder of the 'Cold War' - it is in use as a mobile classroom with a specific purpose as the roofboard indicates:- 'BRITISH TRANSPORT COMMISSION CIVIL DEFENCE SCHOOL'. Freshly out-shopped in lined crimson livery, it was photographed at Bristol in 1957. P.C.

Plate 91. DW150359. CARKIND QPV, is a WR RCE Staff & Dormitory coach which spent most of its Departmental life with Ballast Cleaner DR76218. It was converted in March 1965 at BR Swindon Works from Collett 61ft 'Sunshine' BTK W1610W which was built in 1935 to diagram D.124 on lot No.1574. The ends and the doors in use are in faded yellow, the body is rust-streaked, weathered olive green with white lettering and the roof, (heavily rust stained), and underframe are black. It had a C.3 repair at Doncaster in August 1981 but in 1988 the roof and underframe were badly corroded and the cost of a C.2 repair by 1991/92 was too great and so it was stored at Radyr yard when photographed on 13th May, 1992. It was condemned in 1993 being one of the last four Collett coaches left in Departmental service. S.J.

Plate 92. DW150394 was built at Swindon Works in 1950 and comes from the last batch of coaches ordered by the GWR. It was originally Hawksworth BTK No.W2209W constructed to diagram 133, lot No.1732. Conversion took place in 1967 into a Staff and Dormitory coach to work with the twin jib crane DRP78217 but by 1985, it had been taken out of service. Livery is olive green and whilst its underframe has only had handbrakes added, the window pattern of the body has been changed and the corridor connections have been removed. Photographed at Woodborough in Winter 1969/70. D.L.

Plate 93. DW150395 is a similar coach to the one seen in the plate above and has the same build details, its original number being W2196W. By December 11th, 1992, it was the sole survivor of 18 Hawksworth coaches converted at BR Wolverton Works during 1967/68 into WR RCE Staff & Dormitory Coaches. It was converted in July 1967 and worked with Ballast Cleaners - initially with PWM6360, (later No.DR76208), and later with DR76307 based at Reading. It was lifted and repainted at Cardiff Cathays CWMD in 1988 but subsequently suffered accident damage to an end and was stored at Radyr during 1992. By 1993 it stood condemned at Basford Hall sidings, Crewe and was set aside for Waterman Railways. This photograph, taken at Didcot on 1st October, 1979, shows the other side of this coach compared to that seen in *plate 92* above. B.D.

Plate 94. ADS70126 is a BTU Tool van which was originally built as Maunsell six compartment BTK No.S3676S to diagram 2105 at Eastleigh Works in 1924. It is 59ft long and 8ft ¾ins wide and ran in set 928, (10 coaches), on the Hastings line from 1954 to 1958 but after its withdrawal it was converted in 1963 under the BTU programme. Although the underframe is only slightly altered such as footboard modifications, the body has undergone radical changes with no evidence of its former window and door arrangement. Finished in red livery, it was serving as the Ashford 'Chart Leacon Breakdown Train Unit (Tool)' van when photographed at Stewarts Lane on 12th March, 1978. It was sold from Hoo Junction to Vic Berry of Leicester in November 1990. R.SL.

Plate 95. ADE320388 - CARKIND QQV, was converted to a BTU Tool van in 1956 and served for much of its time at Stewarts Lane where it was photographed on 12th March, 1978. It was eventually sold from Hoo Junction in August 1990 to Booth Roe Metals of Rotherham. Originally built as LNER composite No. 64075 to diagram 205 by the Gloucester Carriage & Wagon Co. Ltd., it became all third No.82445E. There is a slight hint on the bodyside of original panelling but it has suffered radical changes including the provision of end platforms - the nearest one having a lifting gantry. Finished in red livery, it still runs on its original bogies with LNER open front axleboxes. R.SL.

Plate 96. ADM395846 is seen as the Toton BTU Tool van 'MP-M TOTON' and is finished in red with white lettering. It was built as an LMS Period III TK to diagram 1899 on lot No.803 at Wolverton Works in 1934. As E1846M it was converted in 1965 and saw service at Springs Branch and Carlisle Kingmoor. Again the body is heavily modified but the plated over corridor connection can just be seen on the flat end and the roof has retained its torpedo ventilators. Photographed at Toton on 17th September, 1979. R.SL.

Plate 97. ADM395919 was also originally an LMS Period III coach but in this case a BTK numbered M5822M of diagram 1905, lot No.899 which was built at Derby Works in 1936. It was converted in 1965 into a BTU Staff & Tool Coach, (CARKIND QQV here), and is seen in the later yellow and black livery at Buxton on 26th May, 1986. It was based at this TMD and is lettered 'RERAILING COACH/BUXTON TMD'. It is 57ft long over headstocks and has retained its underframe details but the body has been altered, principally beyond the guard's ducket, and the ends have lost their corridor connections. Note the overflow pipes from the water tank in the roof. Except for a short time at Carlisle Currock, it was always allocated to Buxton. A.T.

Plate 98. ADE320912, CARKIND QPV, is a Staff & Dormitory Coach which was converted in November 1961 for service at Lincoln but moved on to Shirebrook, Doncaster and Thornaby where it stood condemned by November 1993. It was one of two, (9011/2), built on order 196 of 1927/28 as an LNER RF to diagram 10C at Doncaster Works. It originally carried number 51773 but was re-numbered to E9012E. There was a first class saloon with 18 seats and a large kitchen area beyond it which was equipped with electric cooking facilities. Some original panelling remains but the large side windows are BR replacements and the ends have had their corridor connections removed. Note how the black stripes on the yellow body are carried a little higher here so that they fit into the central body panel. A photograph of the other side of this vehicle appeared in 'Rail' magazine No. 138 of December/January 1990/91. Photographed here at Thornaby TMD on 27th April, 1988. A.T.

Plate 99. ADE321088 is a 'BREAKDOWN UNIT PACKING VAN' which was converted in 1965 for Healey Mills from BR MkI BS No.M43344 which had been built to diagram 371 (57ft underframe), lot No.30087 at BR York C&W in 1955. Whilst some original double window panels have been retained, the pattern of door and windows seen here is quite different from the original, where there were six compartments each with a door and two windows and a guard's compartment with a single window and three doors. There were no windows in the ends as built. After further service at Holbeck and Knottingley, where it was photographed on 8th February, 1992, it was no longer in use by November 1993. Livery here is yellow and black. A.T.

Plate 100. ADM975161 (should be ADB975161), CARKIND QVX, is a BTU Tool & Generator Van which was converted on lot No.3785 in 1972 to replace DM395382. It served at Cricklewood (1981-1984), Bescot (1985) and then at Allerton (1987-1993) where it was photographed on 31st January, 1993. It was built as BR MkI SK No.M24070 to diagram 146, lot No.3002 at BR Derby Works in 1951. The body has undergone a complete change, the ends have had the corridor connections removed and there have been alterations to the underframe fittings. Bogies and buffers are original. S.J.

Plate 101. ADS227 is branded 'Wimbledon Breakdown Train Unit Tool' (Van) and it was converted in December 1959 from LSWR 'Ironclad' FK No.S7200 built to diagram 476 at Eastleigh Works in 1924. It moved to Clapham Junction CARMD in 1987 but was no longer in use when photographed at Sheffield on 14th October, 1988. Like the conversion seen above, the bodywork has been radically changed and whilst it retains the original buffers, the bogies are replacements - the original type of 'Ironclad' bogies can be seen in *plate 73*. A.T.

Plate 102. KDB975103 is basically the underframe of BR MkI BTK M34705 which was built to diagram 181, lot No.30156 at BR Wolverton Works in 1955. It was converted on lot No.3761 in 1971 and is coded QVV being an S&T cable drum carrier. The added step and handbrake lever are the only notable changes to the underframe. Photographed at Birkenhead in May 1977, it stood condemned in November 1993 at Willesden High Level Sidings. D.L.

Plate 103. ADB975946, CARKIND QVA, is an ER OHLM Cable Drum Carrier which was built as BR MkI SK No. W25243 by Metro-Cammell in 1987 to diagram 147, lot No.30230. Originally fitted with BR B1 bogies, it now has the later Commonwealth pattern. The 'body' is black with white lettering and the cable drum carriers are faded yellow. Note the wheel operated handbrake and the complete all-round handrail which is painted white. The pale blue board (left), is lettered in white:- 'Area Electrification/Manager Doncaster/DO NOT HUMP SHUNT'. It was converted on lot No.3987 in 1980 and was photographed at York Leeman Road Permanent Way Yard on 15th March, 1994. S.J.

Plate 104. This CM & EE OHLM Cable Drum Carrier is one of several which were converted in 1967 from LMS Period III coaches whose underframes were used as shown here. Livery is black with white lettering, handrails, control wheels and steps. Note the lack of side handrails on these early vehicles which would contravene current Health & Safety requirements. Photographed at Longsight OHLM Depot on 1st December, 1979. B.D.

Plate 105. ADB975907. Based on the underframe of BR Mk I CK E15754 built originally by Metro-Cammell in 1955 to diagram 126, lot No.30179, this OHLM Cable Drum Carrier was converted in 1979 on lot No.3972. Originally allocated to Stafford, it later moved to Springs Branch where it was photographed on 30th November, 1996. This end view shows how everything was cut off at floor height leaving the bottom of the gangway connections in situ. The buffer beams are yellow but the other livery details are as for the two previous vehicles. A.T.

Plate 106. ADM395334 bears little resemblance to CR BT No.Sc24286 from which it was converted for the GSU scheme. The underframe, buffers, Fox pattern bogies etc. are the originals but the body has been reduced in height, giving a flat roof for OHLM work and the sides have been horizontally planked. Condemned by 1984 and scrapped soon afterwards, it is seen here in black livery at Wigan on 4th September, 1982. R.SL.

Plate 107. ADE320596 is standing next to the vehicle that was portrayed in the previous plate. This again is a radical conversion. Built as a GER CK to diagram 14600-235, it was later numbered E63645E and was given major body surgery in 1958 when it was converted into a CM &EE OHLM Pantograph Coach. Out of use in peeling olive green livery, it was condemned by 1984. Again much of the underframe is original. R.SL.

Plate 108. ADM395526. This CM & EE OHLM Generator Van was converted in 1959 from the all-steel LMS Period I BG No.M30659M, built to diagram 1715, lot No.344 by the shipbuilders Cammell-Laird in 1928. These coaches had no underframe. Sadly suffering from some vandalism, it was photographed in Rail Blue livery with white lettering when it stood condemned at Thornton Junction on 9th August, 1987. Apart from the lower panel ventilator in one door, it is, externally, as built. It had seen service at Ilford and Glasgow. A.T.

Plate 109. ADM395316 standing next to ADM395526 with which it shares the same build details and livery, this BG was converted in 1958 into a high speed pantograph test coach for the EML electrification scheme. CARKIND QXV, (not carried here), it subsequently followed a similar allocation of duties as the previous coach *(plate 108)*. Note how the roof was changed, the pantograph at the far end, the unusual observation dome and the removal of the guard's ducket. A.T.

Plate 110. LDM395407 is a 'D of M&EE' 'ELECTRIFICATION' Staff coach which was converted in 1958 for the EML. The parent coach was LMS Period I (MR style), TK No.M1214M built to diagram 1756, lot No.7 at Derby Works in 1924. This plate shows the corridor side with the large windows still approximately to the original arrangement but the door pattern has changed. Finished in olive drab with white lettering and yellow and black wasp striped ends, this coach saw widespread service and latterly it was allocated to:- 1983/4 Stratford Market, 1987, Doncaster Hexthorpe ECD. 1990, Millerhill ECD. 1993, Peterborough ECD from where it was condemned and sold to Philips of Llanelli in February 1995. A photograph of the other side appeared in 'Rail' magazine No.212 (October/November 1993). Photographed at Doncaster on 14th November, 1988. A.T.

Plate 111. ADB975701 is a CM & EE OHLM Pantograph Coach allocated to the Birmingham area and based on BR MkI BSK No.M34085, built to diagram 181, lot No.3003 at BR Derby Works in 1951. Conversion took place in 1978 on lot No.3946. The centre and far end doors remain along with four compartment windows but the roof has had the usual changes and the pantograph can be seen at the far end. This is used for test purposes only and does not draw current for propulsion etc. Livery is olive green and it was photographed on 28th July, 1991 at Doncaster awaiting repairs. A.T.

Plate 112. ADB975702. CARKIND QXA, this is a similar vehicle to the one in *plate 111* but we can see the other side here. Direct comparison is difficult because this coach started life as BR Mk I BCK No.E2108 which was built by Metro-Cammell in 1956 to diagram 171, lot No.30185, but its conversion details are similar to those of ADB975701 as its livery. Photographed at Doncaster awaiting repairs on 26th July, 1991. S.J.

Plate 113. LDB975769 was converted on lot No.3955 in 1978 into an OHLM Staff and Office Coach which was based at Romford for much of its time before moving to Doncaster in 1987. Coded QXX, it was formerly BR MkI BSK No.E35251 which was built at BR Wolverton Works in 1957 to diagram 181, lot No.30427. Rationalisation of door numbers and roof alterations are the main external conversion changes along with the fitting of anti-vandal window shields. Livery is again olive green with white lettering, black and yellow 'wasp' ends and a red ladder board. Photographed at Whitemoor Yard on 15th August, 1992.

S.J.

Plate 114. LDM395406 is an Electrification Section Bridge Half Deck Coach which was converted in 1959 for EML from LMS Period I TK No.M1220M, built to diagram 1756, lot No.7 at Derby Works in 1924. The normal height conversions seen so far are too high for overhead line work under bridges, hence the cut down part on this vehicle. After EML work, it was variously based at Stowmarket, Ipswich and Peterborough. Livery as before with white handrails and it was photographed at Whitemoor on 7th April, 1991. After 69 years of service divided almost equally between revenue and departmental, it was sold from Peterborough ECD to Gwent Demolition and Construction in November 1993. S.J.

Plate 115. LDM395328 is an Electrification Section Workshop Coach which was converted in 1957 for the GSU scheme. Originally built at Wolverton Works in 1907 to diagram 133, lot No.B185 as an LNWR TK, latterly numbered M3061M. Coded QYV, but not carried (later becoming QYW), it saw service at Wigan, Barassie and Millerhill where it stands condemned on 16th January, 1993. Livery is the same as in the previous plate. S.J.

Plate 116. ADB975719 is one of nine CARKIND QXX, CM&EE OHLM Stores & Generator vans which were converted on lot No. 3948 in 1978. This vehicle originated as BR MkI BSK E35266 which was built at BR Wolverton Works in 1957 to diagram 181, lot No.30427. This plate shows the compartment side with the luggage compartment nearest. The window pattern is unchanged except in the brake area. Note the handbrake wheel and test pantograph. Livery is olive green with white lettering and roof. Photographed at its base, Carlisle Upperby on 7th December, 1990. S.J.